You Questions - Answered!

Your Path to Becoming an Amateur Radio Operator

by

Christopher E. Cancilla

"**Amateur Radio**" and the original idea of this story are the sole, original, authentic, imaginative, and creative works of **Christopher E. Cancilla**. It is copyright ©2023 by Christopher E. Cancilla and SnipDawg Publishing in all forms, versions, titles, or revisions.

This story, and all published works by Christopher E. Cancilla, can be found at:

http://AuthorCancilla.com

All rights reserved—reprint by permission only.

For reprint permission, send an email to:

SnipDawg.Publishing@gmail.com

with your name, address, phone number, and request.

Be sure to include why you wish to reprint a portion of this text and the section you would like to reprint.

Copyright ©2023-2024

by SnipDawg Publishing

FEBRUARY 2024 –First Printing

Your Questions

→ *Answered* ←

Left blank by design

Thank you to those operators who glanced over this booklet and added or subtracted to the answers. Dedicated and professional people like you give Amateur Radio a good, NO, GREAT name.

My teaming bank of reviewers, editors, and critics is fantastic. I can read something five times, but they read it once and say, "This sounds clumsy!" They are usually correct, also. If they say fix it, I fix it.

What is this book about?

- ***This is a book,*** a collection or compilation of answers to multiple questions and comments by those trying to determine if the amateur radio hobby is something they may be interested in pursuing.

- ***This is a book,*** for the curious person to take a serious look at the hobby of amateur radio with a critical eye, understanding the sport and fun involved with this technical hobby.

- ***This is a book*** that defines the subtext, the shorthand, the acronyms, the rules, and the entertainment value in the hobby of amateur radio. ***HAM speaks***, as it were.

- ***This is a book*** for determining direction. Your direction in a possible new hobby, new fun, new entertainment.

This is a book written for you!

Thank you!

This is where I thank all my friends, editors, and fellow HAMs, but although this is what I will do, I want to also thank all those I consider an EMLER!

An Elmer is a licensed amateur radio operator who takes you under their wing when you start. That was a guy named Elden Morris (N1MN/SK).

Elden got me into the hobby, and I **<u>REALLY</u>** blame him for what I do. Unfortunately, Elden has passed away, but his legacy of teaching and making new HAM radio operators continues.

The /SK after his call sign signifies that he is a *Silent **K**ey*, meaning he died. We will talk of Elden in this book, and although no one can replace him, There is another who I consider an Elmer. He used to live close to me, but he moved. Maybe I scared him. He is N4MQU, Mark Gibson.

Several friends are kind enough to read my work, make corrections, and offer suggestions as to how I can improve the book before you get it in your hands.

➜ INTRODUCTION

Thank you for spending your valuable time reading this book. A book dedicated to and regarding the exciting hobby of amateur radio. Your curiosity about the world of amateur radio has led you here. This book is perfect to see if you will enjoy this hobby.

Many aspects of amateur radio will be described in this text, including analog, digital, FM, voice, CW, SSB, local, worldwide, repeaters, DMR, networks, talk groups, and more. I do not expect you to understand that last sentence.

Still, by the time you finish reading this book, you will understand completely if you were to reread that sentence again. I put it in bold to make it easy to find later.

I have attempted to answer questions thoroughly in this text; if not, please let me know, and I will find the answer and get back to you.

Some have asked me if it is fun, and I say it is. But you will need to decide for yourself if it is fun. It is out of the ordinary. Yes, it can get expensive if you progress and want to get more deeply involved in amateur radio. But you can also spend a limited amount of money and design or build your own equipment, saving you a lot of your hard-earned income.

I know a few people who took part-time jobs to purchase a new radio, and there is me. Unfortunately, I have never purchased a new HF. HTs, yes, but an HF radio, not yet. So, if HF and HT are jibberish, stay with me for a bit. You will understand.

If you think it would be fun and exciting to build your own antenna and connect it to a radio, make a CQ from your home, and chalk up a QSO from someone in another country, the USA, Europe, or Australia, then yes. You would find this a fun-filled and rewarding hobby.

There are other aspects of the hobby. Everyone starts with a Technician class license, predominantly in local communications. Local repeaters, simplex, and perhaps DMR.

Stay with me here; I will define all these terms for you shortly.

Communication is open to you once licensed, even at the Technician level. Any opportunity to speak worldwide is at your fingertips.

DMR, Echolink, 2-meter, 70cm, and limited HF are a few communications mediums available to those with a Technician Class License. The only requirement is passing the FCC Exam.

The way to pass the exam is to study for your exam. Which book makes no real difference. All of the material in the book is the same. It is how it is presented in the book that makes them unique.

Is one better than another? Not really. Some students told me that one book is better than another and vice versa. Pick a book, read it, and repeat.

So then:

1. Find a book,
2. order it,
3. buy it, and
4. read it.
5. Then go to **HamTestPrep.com** or **HamStudy.org** and take practice exams to see what you are retaining and what you need to study deeper.
    a. Register on your chosen site, which tracks your progress and gives you weaker subjects more often.
    b. There are 750 possible questions.
    c. If you see them all when taking practice tests, then you see them all. There is a better chance you will see and know/recognize those questions on your FCC exam again.

In the next section, we will discuss specifics. So stay with me here; it will clear as you read, and you may not even realize it.

➔ GLOSSARY #1

HT	Handheld Transceiver: please don't refer to it as a walkie-talkie.
HF	High Frequency, used for worldwide radio communications.
CW	Code Work, essentially Morse code communications.
QSO	A conversation over the radio.
DMR	Digital Mobile Radio.
SSB	Single Side Band is used as a mode in HF communications.
CQ	A call out to talk to anyone, as in "CQ, CQ, CQ," you want to speak to anyone worldwide.
DX	Distance Communications, as in "CQ DX, CQ DX," You want to talk to anyone not in your country.
QSL	Confirm Communications. Some HAMs send a QSL card after a QSO.
73	HAM Speak for My best regards. Not Seventy-three, or seven-three's. Seven – three.
USB	Upper Side Band, SSB protocol is used when the frequency is above 10 MHz (below 20-meter and band).
LSB	Lower Side Band, SSB protocol is used when the frequency is below 10 MHz (40-meter band and above).
SSB	Single Side Band is typical communication when communicating on high frequencies.
CW	Carrier Wave is also known as Morse Code or Code Work.
Call Sign	A specific ID or name granted to an operator in the USA by the FCC
FCC	Federal Communications Commission

➔ REPEATERS

Local repeaters will allow your little 5-watt HT to communicate with someone a hundred miles away. It does this by receiving your transmission and retransmitting it with a much higher power level and from a much higher altitude. When it retransmits your voice, it does so on a slightly different frequency. This is known as the offset frequency.

For example, you transmit to the repeater on 145.600 MHz (MegaHertz or MHz). The repeater recognizes your communication and retransmits your voice at 145.000 MHz. This is a positive offset for the repeater. Everyone listening to that repeater listens on 145.000 MHz. So, everyone within reach of that repeater can hear your call. When you release the PTT (Push To Talk) button to listen, your radio listens to the repeater frequency, which is 145.000 MHz. It is automatic. When you press the PTT button, the automatic switching in your HT (Handy Talkie or Handheld Transceiver) transmits on the offset frequency, which can be a + or a – (plus or minus) offset. Hence, when you transmit, the radio switches from 145.000 to 145.600. As I said, this is referred to as the repeater offset. For 2 meters, in a 144 to 148 MHz frequency range, the offset is .6 MHz or 600 kilohertz.

If you use a 70 cm (centimeter) repeater in the 440 MHz area, the offset will generally be 5 MHz. This can also be a + or a – (plus or minus) offset. The offset comes into the repeater frequency when you transmit. The "base" frequency is always the repeater's listening frequency.

There is one more aspect of a repeater you need to know about. It is called the Tone, the PL (Motorola name), Channel Guard (GE name), or simply the CTCSS code. This is an acronym for **C**ontinuous **T**one-**C**oded **S**quelch **S**ystem. This is a sub-audible tone transmitted to the repeater from your radio to let the repeater know that I am trying to communicate my voice through you. It falls just below the range of human hearing. Using a CTCSS does not mean it is a CLOSED or PRIVATE repeater, but rather that the repeater owner avoids interference that may or may not be present from a nearby or distant repeater. The repeater owner determines the CTCSS tone to use or not to use if they choose.

When the CTCSS tone is received by the repeater, it opens up. It transfers your communication from the RECEIVE side of the repeater to the TRANSMIT side. Your voice is being retransmitted in real-time to all listening as you are speaking. PL is a Motorola-specific name, meaning **P**rivate **L**ine. They are the same thing.

Repeater connection information would be written as:

145.000 (+) PL: 88.5

This means you listen to the repeater on 145.000 MHz. It has a plus offset, so you transmit to the repeater on 145.600 MHz. Finally, you set the CTCSS or PL to 88.5 Hz to "open" the repeater. This allows your voice to move to the transmitter, which transmits over the base frequency as you speak. Simple right?

The footprint of a repeater depends on multiple things. Two of which are the antenna's height and the repeater's transmit power level. The higher the antenna is on the tower, the farther it can reach. This is due to covering more of the Earth's surface from a greater height. That pesky curve of the Earth's surface limits the line of sight.

To give you an idea, two people on 5-watt HTs talking will have an effective range of fewer than 6 miles if there are no obstacles on a flat surface. Believe it or not, this uses the formula (in meters): antenna height x horizon at ground level = antenna horizon. Therefore, 1.8 meters (5 feet) x 3.6 kilometers (horizon at the surface) = about 6.5 kilometers or maybe 4 miles.

When I say surface, I mean the location of the horizon as if you are lying on the ground, not 5 feet above it.

This assumes the transmitting antenna is 5 feet off the ground and the receiving antenna is lying on the ground. Therefore, if both are 5 feet off the surface, the expected range would be 13 kilometers or 8 miles. The higher the antenna, the farther the distance you can communicate.

If you want to play with it, I plugged the formula into MS Excel.

Antenna Height (feet)	Antenna Height (meters)		Repeater Range
	1.8	=	4.788
9.0		=	5.956
1200		=	68.771
	366	=	68.279
	3	=	6.182

Formula: `=IF(B3>0,SQRT(B3)*3.569,IF(A3>0,SQRT(A3/3.281)*3.596))`

If you like playing with math as I do, this will give you something to do for a while.

Using the same formula at 1,200 feet, like the K4ITL repeater in Auburn, NC, the repeater at 5 watts can transmit about 70 miles on paper. However, repeaters transmit a greater power than an HT, 50 to 150 watts or more. This will give it a better reach, of course. I know on K4ITL that I have spoken to someone at the Outer Banks of NC, 160 miles from the repeater's transmitter. I was 40 miles in the other direction. This means I could talk to someone 200 miles from me on my little HT. That is what a repeater is for!

Another exciting use for a repeater is circumventing an obstacle, like a mountain. For example, Don and Doug live on opposite sides of a mountain, which means that on a simplex frequency, they cannot get a signal to or from each other. Therefore, the transmitted signal will not pass through the mountain.

Don purchases a repeater and places it on the mountain to negate this minor issue. Now, they can both hit that repeater, so they have excellent communications without considering the mountain because they found a way around it! Well, not around it, but over it for sure. How tall the mountain is and the distance between them disappears.

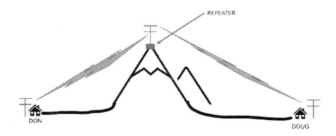

A repeater is powered by electricity. So, the repeater needs an electric POWER source to run. But it can use DC power. This is what a battery will produce. So, a large enough solar panel and a few deep-cycle marine batteries connected to the power input will allow this repeater to work forever! As long as it gets sun during the day to recharge the batteries.

The last thing about a repeater you should know is that a repeater can be connected to another repeater, or an entire repeater network, to provide communications a great distance or even on or over multiple mountains. This is accomplished by connecting the repeaters, the transmitting, and the receiving, to the Internet. When you speak to the repeater, your voice is being retransmitted through the repeater you are talking to. Still, it is also being sent to all other repeaters on its network through the Internet. So when you let go of the PTT button and someone responds, they may be hundreds of miles from you. Still, because you are both close to your respective repeaters, communications are crystal clear. As we say, you have full quieting.

Now that the term REPEATER has been beaten to death. Let's talk about another you may need to know.

→ SIMPLEX

Simplex:

Who remembers RADIO SHACK and their walkie-talkies. Tuning to channel 19 and talking to truck drivers. Well, this ain't it!

Simplex is the same as buying those blister pack radios in the big box stores but with a more robust and better-sounding rig in your hand. By the way, those blister-pack radios vary in pricing quite a bit. The FCC authorized the newer radios to transmit at 2 watts, increasing from the previous 300 mw. This really means a greater distance but, more so, clearer communication. Remember we talked about height above the ground already. Also, their mode is FM, like any amateur HT, instead of AM, like a CB radio. This clears up the voice quite a bit. Look for a quality unit, don't get sucked in by the MILES it can talk. They can all talk the max distance if they transmit using two watts of power. Stick with a name brand you trust. You can usually get a couple of them for less than $100.

Can you talk 68 miles like the package says? Yes, you can. If you are both at the top of a mountain, in clear view of each other, with minimal atmosphere and no interference.

Can you talk a mile or two? They are designed to be used in buildings and the woods; 70 cm is perfect, so you should be OK if you are within a few miles of each other under optimum conditions.

The usual range of an FRS device on channels 8-14 is less than one-half mile, but longer-range communications can be achieved on channels 1-7 and 15-22, depending on current conditions.

I also advise getting the NOAA channels on the radio if you go the FRS route. Great in an emergency or to keep tabs on the weather where you are located. Some can be set to blast an ALERT for intense weather like a tornado. Consider this a notable safety feature. It can be set to listen to the emergency and sound an alert. I used to put one radio on the alert setting, put it in a Ziploc Freezer bag (waterproof), and tie it to a tree in the campsite. Then, when the alert told us the weather was about to go south, someone heard it and started the emergency process for the group.

I have had people tell me, "Hey! I picked up some Chinese radios and programmed them for NOAA and FRS." My response is simple, "Hey! If the FCC catches you, they can fine you up to $10,000 per incident when, not if, they catch you." Their response is typical, "Well if that happens, it's better to ask forgiveness than permission." I smile and tell them the cost of forgiveness is a court case and a $10,000 fine that they need to pay. Some people think it's OK. All I can do is tell them the

WHYs and the WHAT FORs. They need to determine if playing in the legal area is for them.

Another technician class amateur radio technology available to a licensed person is DMR. It is a technology requiring you to pass the FCC exam for the Technician level at a minimum. It is a **D**igital **M**obile **R**adio format that permits two radios to send digital packets to each other, the packet containing your voice.

As mentioned in the previous text, you can send these digital packets directly to another radio (like simplex) or a digital repeater. But with DMR, you can also connect through a hotspot. A small device connected to the Internet allows your HT to communicate worldwide, opening up the entire world as a possible group that can talk with each other in real time from anywhere.

I have discussed becoming a licensed amateur radio operator with several people. Their issue with licensing is surrounded by the view they cannot learn the information and pass the FCC exam. This is false. Anyone can learn the information and pass the exam.

For example, look at the 9-year-old young lady who is licensed and was given Technician at the same time as I was, but with a better score. Yeah! Never lived that down, but she had fun taunting me. She was a little pointed to make sure I knew, through her father, that she passed her General about 5 years later, the second level of licensing, a week before I did. But I passed Extra a few weeks before she did, and I told her Dad to tell her I made Extra before her. He laughed. After all, Dad is a General (he has since passed Extra). She is a 16-year-old Amateur Extra class operator with hundreds of QSOs more than me. It's easier for her since they have a complete station in their home. A QSO is contact with someone over the radio.

I need to set up a time for an HF QSO with them.... But I get ahead of myself.

Before you begin reading about the various levels of licensing, a bit about me.

→ How did this radio thing all start?

I was the Scoutmaster of BSA Troop 723 in Powder Springs, Georgia. A most excellent friend (Yes….. a Bill and Ted reference), Elden Morris (N1MN/SK), told me he would hold a Technician Licensing class on three consecutive Saturdays in December 2013. Anyone who attended the course could take the FCC exam; if they passed, they would be licensed. In addition, those Scouts who attended and actively participated would complete all the requirements for the Radio Merit Badge.

If you are unaware, the /SK after a call sign designates an operator who has passed away. It stands for (S)ilent (K)ey.

But before I met Elden, I knew Roger. I have been interested in amateur radio since 1980 when I had a roommate in the barracks in the Air Force who was an Amateur Extra (NJ7T), had a radio in the room, and talked all over the planet. So his name is Roger (NJ7T). But at the time, I had a problem learning Morse code. So, I think he let me play with his Yaesu FT-101 (101BF?). We put up an antenna outside the barracks and had a blast.

I thought it was very cool! I attempted to get my license back then but did not pass the Morse code portion of the exam. Since those days, the FCC has removed the requirement for Morse code. So, you can become a licensed amateur radio operator if you pass the written test.

Back to my story, I passed the FCC Technician Exam and became licensed in December 2013. However, Elden started the next day, actually that same afternoon. Wait, it was as I turned in my technician exam….. Anyway, he started asking me when I would take the General exam. Unfortunately, I did not understand what it meant to be a Technician class operator yet and kept putting it off.

I bought a $25 HT (Handy Talkie) – Baofeng UV-5R. I learned to program and use it, always had it on, and listened to the repeater, W4BTI, in Atlanta, but I was very mic-shy and never said anything. Then, one afternoon, I heard someone call for my call sign; OK, now I had to answer. Yep, good guess. Elden called me to get me on the air and overcome my shyness about speaking on the radio. We were talking on the phone as he was driving home from somewhere, and a few seconds after we hung up, he called for me on the repeater. He managed to get me comfortable and took me to a club meeting. I joined the Kennehoochee Amateur Radio Club (W4BTI) in Atlanta that night. I did have fun and lived on the repeaters in the area and when traveling in my car.

For years, all I had was my trusty HT. Still, I set up an external antenna to make communicating easier when driving. It was a magnetic mount antenna on top of my car and a little adapter to connect the cable to the radio. It worked like a champ.

Elden showed me I could get a Vanity Call Sign, so I applied for and was granted W4CEC. It replaced my original call sign of KK4VQP. He also showed me the QRZ.com website and told me I needed to claim my call sign and put information there since now that I am 'on the air,' people will be looking me up.

On my QRZ page, I made sure to blame **Roger** and **Elden** for my involvement in Amateur Radio *and thanked them for pushing me*. I really owe a lot to Elden; he was a great example, mentor, and the best ELMER any new HAM could hope for. I hope to continue his passion for helping others learn and enjoy this fantastic hobby. **He was my first Elmer**.

An Elmer is an older HAM who 'shows you the ropes.' Things like antenna design, radio info, and everything you need to know to be an Elmer yourself one day. Elden's wife always told him she did not want to be licensed because she did not do public speaking. My wife is the same way. She supports me in the hobby but has no interest in getting licensed, and I understand.

In June 2017, I moved back to Raleigh, North Carolina. After a few months, I met the Franklin County Amateur Radio Club (AA4RV). I got a taste of a General class operator's radio life and activities through the summer field day, and it was a lot of fun. So I studied for the General exam, passed it in July of 2018, met up at the American Legion with Ken (KN4GUS), and had to exercise my less-than-one-hour-old General license. So, W4CEC/AG made contact on Ken's radio with a guy in Texas, K5TR. It was terrific, and I found it to be addictive! Thank you, George Fremin, for being my VERY FIRST HF contact, with a less than 1-hour old General license using a /AG on my call.

I had so much fun as a General that I studied for the Extra exam and passed it at the end of December 2018.

My new local Elmer became a guy named Mark Gibson (N4MQU). He was not an Elmer in the traditional sense but in the TEACHING and TESTING sense. He was not one to experiment with antenna design, but he used all the antennas he acquired.

I took the extra exam and passed. I was not fully aware of it then, but Mark planned for me to be a VE, a Volunteer Examiner. The person who provides the FCC exams. He did not stop there, though. He talked me into creating my VE Team and running exams in the Wake Forest, NC, area.

How he did that, I still have no clue!

But I hold between 3 and 6 licensing classes a year, mainly at the American Legion building, where my wife and I are legion members.

Mark's goal was to make me a team lead for a VE team. I would take his "education" and start a VE team, providing the FCC exam at **zero cost** through the Laurel VEC.

I called Elden, and we discussed it, and he told me to do it and listen to Mark. I listened to Elden; therefore, he told me to listen to Mark, so I did.

Several organizations can offer FCC exams. They are called VECs, which stands for Volunteer Examiner Coordinators. These organizations can create teams to proctor the FCC Amateur Radio Exams. I am a VE for the ARRL and Laurel VECs. In addition, I am the Team Lead for LaurelVEC, W4CECVE Team. You can check out the Laurel VEC at **https://www.laurelvec.com/**

A good friend, Kevin Carpenter (KB4KAC), passed General the same day I did. It was odd. Neither of us knew the other was even a holder of a Technician license. However, I managed to draft Kevin into my VE Team, and he is the #2 guy, my Deputy, on the team. So never fear, all of you; Kevin found a way to get me back. Read on!

My other #2 is John Deacon, KB4OI. Kevin and John are interchangeable, but I think John is a little taller. The two of them are a lot better at the non-phone (voice) side of using HF, as in digital connectivity and communications like WinLink, FT8, and others. Most of the time, I defer to their knowledge. If I say something wrong, trust me, they will let me know.

John is the current president of the Franklin County Amateur Radio Club. Between the three primary clubs, we hold exam sessions several times yearly at the American Legion in Wake Forest, North Carolina, for Legion members, Scouts and Scouters, and the local community.

By the three primary clubs, I am referring to the TALARC, the OARS, and the FCARC.

TALARC is **T**he **A**merican **L**egion **A**mateur **R**adio **C**lub.

OARS is the **O**cconeechee **A**mateur **R**adio **S**ociety.

FCARC is the **F**ranklin **C**ounty **A**mateur **R**adio Club.

We average between 12 and 50 test-takers in the class/exam, and coordinating can be a challenge, but it always works out. As a contractor/consultant, I learned that you never apologize or let on the event or the meeting or whatever is not going as planned, at least to those in the meeting, or in this case, the exam. Those people will not be aware there is a plan B, C, or L in effect. They will simply sit, take their test, pass, and leave thinking, **"Wow, that was well organized!"**

Other classes and exams are held at the Scout Camp in Carthage, NC, and usually at another location as needed.

The OARS club tries to have at least 2 to 4 Radio Merit Badge classes each year, and Kevin and John are instrumental in the success of those classes.

Once a year, I attend SWC (**https://SWCbsa.org**), and this year, Amateur Radio plays a more prominent part. We will have dedicated frequencies for licensed operators, and we will also have a Test Session on Saturday afternoon around 3:30pm. The Course Director and a few others are studying for the exam now. I hope they pass!!

To hold a test session, we only need three Amateur Extra VEs and people willing to take the exam. If they pass, the FCC awards them a call sign. If only Technician tests are being taken, a General can be the VE, but 90% of all Techs that pass also accept the General exam, and about a third pass the General. To grade a General exam, you must be an Extra Class VE.

I also started a TALARC (post #187) club. (**T**)he (**A**)merican (**L**)egion (**A**)mateur (**R**)adio (**C**)lub. Our "thing" is training, education, and community service. Kevin passed Extra a short time after I did. As I mentioned, he is much better at HF (high frequency) digital than I am because he has done much more. He is in a house with his HF radio and multiple antennas (antenna farm?) on his property. I am in an apartment, so my HF use is limited to when I am not at the complex. I prefer POTA activations or Parks on the Air. But when at that Scout Camp, I do some POTA Hunting. That is when you do nothing but make contact with operators who are activating parks.

Check out **https://POTA.app** for more information.

LOGO ON THE NEXT PAGE

OK, Here it is............

Kevin Got Me Back – But I love it!!

Never fear; Kevin got me back for drafting him into my VE Team and making him my Deputy. After that, he called for me to become a CAI (Certified Angling Instructor). We teach the Fishing Merit Badge and help Scouts of all ages learn about fishing. So, I spent many Wednesdays learning about fishing, teaching fishing, and identifying fishing......I mean identifying various fish.

Back to my road to Radio

Since that January 2019 day, my first exam session, our group has planned and held two Technician Licensing Classes (like I took from Elden all those years ago) and one General class each year. We also hold at least three Radio Merit Badge classes a year. Still, any Scout who attends and participates in the Technician Licensing Class also completes the Merit Badge. Also, I'll have a conversation with any Scout who is a licensed amateur operator to ensure they understand the information and award them the radio merit badge if they do not already have it. You can wear the Amateur Radio operator patch on your right sleeve if you are a licensed amateur radio operator in Scouting, youth, or adult. That is the only requirement. I give them to the person at no cost.

Amateur Radio Operator

Each year, we also hold a Technician licensing class and two radio merit badge classes at Camp Durant in Carthage, North Carolina. We hold them at the outdoor shelter at the Health Lodge on the camp and always camp in the adjacent campsite.

At this writing, we have licensed just under 250 people aged 10 to 80. In addition, I have signed more than 125 RADIO Merit Badge Blue Cards. Thankfully, nowadays, merit badge sign-off is all online.

During the classes, we discuss the material, of course. But we also take time out to discuss the equipment. What do you want versus what do you need to get started. If you take advice from people who have been in the hobby all their lives, the cash outlay can slope upward at a very sharp angle. You can get started for about $60 with just an analog radio. Still, if you have $300 lying around or available, I suggest you get a digital radio for your first radio; it has more capabilities and is built better. Check out the Anytone 878UV2 or the Btech 6x2 Pro. Same radio, essentially, and I love mine.

Having a mobile radio in your car with an antenna magnetically attached to your roof will significantly enhance your capabilities to talk to someone or a distant repeater. But honestly, a magnetic antenna and an adapter to connect the cable to your HT is an excellent start for less than $100; even $50 would get you something good. That would be the antenna and the adapter to connect to the radio. I used that for over five years until I got my mobile in my car.

➔ What about equipment?

I currently have multiple Baofeng and Wouxan HTs but primarily use my Btech 6x2 DMR. I have given most of the Baofeng radios to new technicians, one of whom is my son's friend. Also gave him a connector and a mag-mount antenna. Man, that brought back some memories!

I have the cutest little radio I picked up at the RARSfest.

(This is a HAM fest in Raleigh).

It is an Alinco DJ-C5. It is the thickness of a few credit cards and similar dimensions.

Unfortunately, it only has a 300mw output, so long-range is not in its vocabulary. It is, however, perfect for very lightweight carry when communications are for close-by receivers.

As for an HF radio, that would be my Yaesu FT-891. I use this for portable operations, like setting up at a Scout event or in a park. Recently, I have been doing a lot of POTA activations; more on this later.

If not at the house, I run off battery power, a compact 6Ah LiFePo4 battery pack weighing 1.6 pounds or my 70Ah SLA (Sealed Lead Acid) battery weighing 60 pounds. I plan to install a solar charging system to remain on the air indefinitely.

I picked up a small (12x12 inch) solar panel and charge controller with 200 watts on it, and I am skeptical it is truly 200W because of its size, but hey, never know. It is for ONLY a cell (5v) charger, but it may be handy somewhere.

My antenna is a Wolf River Coil SB-1000TIA. I had a dual collar attached at one point. This means I can set one collar to 40 meters and one to 20 meters. Then, to switch between them, I only need to move a wire, a 10-second band change! I found tuning it with a single collar much more manageable, so I removed the second one from the coil.

I did add one more thing to this antenna: a longer antenna. It uses a collapsable whip on top of the coil. The original whip is 102 inches, but I recently picked up a 204-inch whip, and I get better performance when I use it.

Tuners help the antenna be close to perfect. SWR, or Standing Wave Ratio, is the amount of output power from your radio reflected from your antenna back into your radio. A 4.1 to 1 SWR (4.1:1) is the most you want to have. I try to get as low as possible; I am pretty happy if the lowest possible is 2.5:1.

Suppose the best your antenna can do is 5.5:1. In that case, a tuner can bring it to a 1.1:1, which is a perfect match, perfect SWR. All the power goes out through your antenna when it is a 1.1:1, ensuring you will be heard clearly. I found a used MFJ-939, but I also picked up an ATU-100. Either of which can auto-tune the radio. But remember one thing. The tuner fakes your radio to believe the SWR is a 1.1:1 match when it can be as high as 15:1. THAT IS REALLY BAD. When you transmit, the coil is heated because of the inefficient matching between the radio and the antenna. Still, the tuner between the two is compensating for the horrible SWR. It works, yes. But is this the best answer? NO!

Tune your antenna as close as you can get, then if you want a better tune, use a tuner. Your radio will be happy. Just keep an eye on your antenna so it is not superheated.

I use QRZ most for logging contacts and LoTW, the world's logbook. The two are compatible, so I can export from one and import to the other. I also use L4OM, which would be LOG for Old Men, and yes, I qualify.

I have several other radios, but I play mainly with the ones mentioned above. My main HT is the Btech 6x2 DMR; I have it set up for DMR (I have a hotspot in my house) and all my analog communications. Both simplex and repeater. Locally, I have used a few digital repeaters.

Someday, I hope to learn and begin using CW, also known as Morse Code. CW stands for Continuous Wave. Some think it stands for Code Work. Whatever, it is just CW, and I need to learn the code before I am useful.

There are three licensing levels for an amateur operator. Technician, General, and Amateur Extra.

Each is designated for specific bands and/or frequencies, with Extra having full FCC privileges. What does that mean? The United States FCC band plan shows those privileges at a given band in the image below.

The chart below is from the ARRL; you can download it and print it out or purchase copies in various sizes. If you are a general or Extra, you need a band plan with your radio when turned on. This will ensure you are within the frequencies of your license limitations.

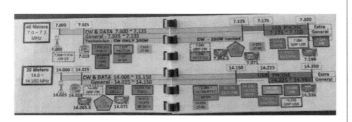

I keep a band plan booklet I bought from Silver Tip Antennas when I operate. You can get your copy at:

https://www.qrz.com/db/KJ4ADN

There are still HAMs with a Novice or Advanced license, and this chart reflects their usage areas. No one can be a new Novice or Advanced, only a **Technician, General**, or **Amateur Extra**.

Let's take a look at the 10-meter band as an example.

We will ignore the RED areas (data) and focus on the GREEN and YELLOW in this 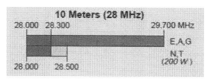 image. As you can see, the entire band is available for phone (voice) for the E(extra), A(advanced), and G(General). Phone means voice and typically uses SSB, Single Side Band. In addition, there is an Upper Side Band and a Lower Side Band.

When using SSB for voice communications, you have two options. The general rule of thumb is to use the Upper Side Band, USB, on frequencies above 10 MHz and the Lower Side Band, LSB, on frequencies below 10 MHz.

On 10 meters, the frequency span for the available band (for the extra class operator) is 28.3 MHz to 29.7 MHz. Therefore, for the 10-meter band, the generally accepted SSB setting is USB or Upper Side Band.

In the yellow portion, you see N and T, Novice and Technician, are also available. However, the FCC limits a novice or Technician from 28.3 MHz to 28.5 MHz and effectual power output to 200 Watts. The other classes are limited to the maximum power permissible by the FCC, 1,500 Watts.

Is it worth getting an amplifier? Yes, if you need to break through a pile-up and be heard. But the cost is tremendous. A good guess is about $2 to $3 per watt, so a

500-watt amplifier will cost between $1,000 and $1,500. You can purchase used equipment, and HAMs take good care of their toys, which is a real option. All of my HF radios are used. I have never purchased a new HF radio, as the total NEW price scares me!

I have never had more than a bare-bones output, 100 Watts (150 Watts on the FT-900), and I can usually make contact quickly. However, some who have 100 Watts and lower employ the slip method (that's what I call it) to break through many people attempting to contact a CQ. This is when you get on the frequency and have an issue being heard.

Then, after a few traditional calls, you de-tune your radio a few kHz one way or the other. This makes your voice sound odd and unique, making you stand out. The person calling CQ, the one you are trying to reach, will hear you a little better because of your uniqueness.

Once you call, reset to the primary frequency and make the QSO as usual.

➔ GLOSSARY #2

I just used a few unique terms: CQ and QSO. What do they mean?

CQ means, "Hey, everyone, I'm here and want to talk to anyone."

CQDX means, "Hey, everyone, I'm here and want to talk to anyone <u>NOT</u> in the same country as I am."

A typical CQ call could be, "CQ CQ CQ, W4CEC, CQ CQ CQ" or you could say Whiskey-Four-Charlie-Echo-Charlie as a phonetic way to make your call sign easier to hear and understand.

That is because if you are low, in the static, or drowned out for some reason, the phonetic will give the receiver the chance to hear your call clearer. For example, the letters C, D, E, G, P, T, V, and Z can all sound similar if there is static or the receiver has low audio. Using the phonetic terms for the letters dispels second-guessing and improves accuracy.

No guessing, so they become Charlie-Delta-Echo-Golf-Papa-Tango-Victor-Zulu.

Those are standard, but you can use Carrot and Dandy or any other term that comes to mind. I have used a few interesting ones and always got my message across. The problems arise with a purist. If, for the letter E, I were to use EGGPLANT instead of ECHO, some people out there would scold me rather than accept the phonetics I am using. Do I really want to talk to those people? Well, maybe? But at the end of our QSO, I may have been known to say......

"73, and thanks for the QSO. This is Water-4-Chocolate-Easy-Chocolate clear."

I could feel them cringe as they signed off.

Another question you may have is how they know if I am in their country. Simple, my call sign. Every country has a designated call sign 'type' or prefix. For example, A, K, N, and W are the norm in the United States, as only in the USA can a call sign start with those letters. VE or VO in Canada could begin a call sign. KL is Alaska, KP is Puerto Roco, and KH is Hawaii. And each has its own.

Here is an image to help you visualize.

A QSO is nothing more than a contact, a conversation. At a minimum, a hello! Which can last as short as a few seconds if it is a contest or a ragchew if you are simply chatting. A ragchew is back-and-forth communication with someone else or several others on a specific frequency. They can all be in the same country or not.

One fun activity in HF is POTA and/or SOTA. This is **P**arks **O**n **T**he **A**ir and **S**ummits **O**n **T**he **A**ir. Activating a park is fun. You put your call sign there as a POTA activator, and they flock to you. Make 10 contacts, and you just activated that park. Same for Summits or mountains. It is all in good fun but simultaneously allows you to increase your speed in setup, making contacts, and ultimately changing bands and frequencies. I call it your radio muscle memory. Do it enough, and it requires no thought to complete the task.

There is one other type of QSO, the EYEBALL QSO. This is where you and the other HAM operator are standing in the same room, as in eye-to-eye.

You also need to know a few other terms: QRM, QRN, QRP, QST, QSL, QRY, QTH, QRZ, YL, and XYL.

Here is what they are; they should be pretty easy to understand.

QRM	Man-Made Noise
QRN	Natural Noise
QRP	Low Power transceiver/communication
QRT	Shutting Down
QSL	Acknowledge(d)
QSY	Changing frequency, changing band
QTH	Home location as in city/state or Grid Square
QRZ	Who am I talking to?
YL	Young Lady
XYL	Former or Previously a young lady (wife)

The last two are still used; I have heard an OM code. Yes, it stands for Old Man. So, in that light, one of my HF contact logging programs is called LOG-4-OLD-MEN, or L4OM. I just got turned on to a new logging program I use on my PC and phone. It is called HAMRS (Hammers). I use this now for almost everything, especially my POTA (Parks on the Air) activations. It is simple and easy. It also directly interfaces with many other online logging programs, so uploading my logbook is a snap.

➔ CALLSIGNS

Speaking of call signs a moment ago, I thought I would explain them more in-depth. My call sign is **W4CEC**. Does that mean anything? In a way, yes, it does.

It is a vanity call sign, meaning I petitioned the FCC to change my call sign to this permanently. Why, well, CEC are my initials.

You see a W, the number 4, then my initials, looking at the call sign.

The 4 designates what is commonly called 4-land, or the Southeast United States.

There are also limitations in the call sign itself. Call signs are referred to as a 1x2 or a 1x3. Mine is a 1x3. 1 letter, the number, 3 letters, so 1 and 3, or a 1 by 3. I got this call sign as a Technician and decided to keep it forever. However, some people change their call sign on a whim. I am not one of them.

A 1x1 (N4C) is only for a Special Event station and a limited number of days. I always try to reserve N4C for Jamboree on the Air in October.

A 1x2, 2x1, or 2x2 (ex. W4CC, WB4C, WB4CC) call sign type is only available to an Extra Class operator. You cannot get one of these if you are a general or a technician. Once you pass the extra exam and the license is in the ULS, you can do one of two things. Request the FCC assign the next Extra Class call in sequence at that time, or look for, find, and request a specific call sign as a Vanity call.

I grew up in Ohio, so W8CEC would have been fine. However, W8CEC is the license of Chad E Carter, and I am sure he is thrilled with that call sign. I live in North Carolina, and I found W4CEC available. When I got that call sign, I lived in Georgia; I applied, and it was granted. DONE DEAL!

In the USA, a call sign can begin with:

- AA through AL
- K
- KA through KZ
- N
- NA through NZ
- W
- WA through WZ

The map below shows you the *'land'* for each number.

Your first call sign that you will be issued by the FCC will be the next available call sign in the current sequence. I believe they are up to a KQ or KR call now, and once they run through all the end letters, like AAA, AAB, AAC, and AAD through ZZZ, they will move on to KS call signs.

Suppose you are well-studied, as in taking the Technician, General, and the Extra exam in one sitting and passing all three tests. In that case, the FCC will issue you a 2x2 call sign since you will be an Extra Class Operator.

➔ CASH OUTLAY

Now, we have discussed call signs, antennas, radios, and repeaters. Then, we talked a little about NOAA and FRS.

One crucial thing to mention next, I am sure you will agree, is the **CO$T** of getting into this hobby.

Let's see:

1. You need to purchase a book to aid you in learning the material
    a. $25
2. You may want to take a class to assist you in understanding the material you are studying
    a. FREE – check out https://W4CEC.com/registration
3. You need to take the FCC exam
    a. Laurel VEC, FREE
    b. Other VECs are $10 to $25 per exam session
4. Pay the FCC FEE
    a. $35
        i. This is only required for the following:
            1. a NEW license
            2. request for a VANITY call sign
            3. RENEWAL for a current call sign
        ii. There is NO FEE for upgrades (Technician to General or General to Extra). If you are brilliant and go from Technician to Extra in one

test session, your total cost would be $25 because you hit the trifecta, as in passing all three exams in one test session. You came in with no license and left with Extra!
5. Purchase your first HT
    a. $25 to $90 for a basic analog HT
    b. $200 to $300 for a DMR HT
        i. Depends on what you are looking to do
        ii. For less money and analog-only, I really recommend a UV-82C or even the UV-82HP (about $60)
        iii. For more money and analog and digital operation, I recommend either the Btech 6x2 Pro or the Anytone 878UVIII (about $250 to $350)
        iv. You will be much further ahead with the DMR (digital) radio.
6. Purchase your first HF
    a. USED: $200 to $1000 or more
        i. This depends on what you want to buy
    b. NEW: $400 to $5000
        i. Depends on your expendable income, I suppose
        ii. Or, if you have an understanding spouse
7. Antenna
    a. Free to $$$$$
        i. Have stuff lying around in the garage. Make your own.
        ii. Portable, Base, Mobile
        iii. Low, medium, high quality

        iv. Off-brand vs. respected brand name
        v. You really need to ALSO have an SWR meter
            1. I have a NanoVNA, which is about $100, which makes for a great SWR meter and many other things.

I own several HTs, a couple HFs, and a few antennas. Sometimes, I get bored and go to the garage and make a new antenna, test it, hook it up, and give it a try. For 2-meter I can whip out a slim jim in less than an hour and give it away to people. I got the window wire from a friend with a roll of it.

I keep my radios, connectors, SWR meter, roll-up antennas, and my complete HF rig in waterproof toolboxes filled with foam. As a result, I can take a truly mobile and portable GO KIT anywhere, set it up, and get on the air in less than an hour.

The oddest antenna I have built is a Slot Cube antenna for 2-meter operations.

Yes, that is all made from copper water pipe, you need to solder it, and it is all connected to a PVC pipe. Took a couple hours to assemble. Tested it, and the SWR was pretty low, 1.5:1 over the entire band. Hooked it to a radio and did a test, which worked well. Has interesting propagation and gain, I estimate.

Since I used a low-power HT, it caused the cable box to scramble when I pressed the PTT button. Tells you how good the cable system is in this apartment complex. You cannot wilfully cause QRM (remember what that is, right?). **QRM is Man-made Interference/noise!**

Since I am fully aware that the antenna causes this issue in my apartment complex, in just 5 seconds of transmitting, I cannot use this antenna at or near my apartment. As I said a moment ago, **you cannot wilfully cause QRM**.

A moment ago, I mentioned a DMR, Digital Mobile Radio. Yes, you need a call sign, but more importantly, you need a digital ID. Once your call is listed with the FCC, you can apply for a digital ID at **https://RadioID.net**, and in a day or two, you will receive an email with your brand-new digital ID. Consider it a digital call sign. If you get a DMR radio, you will need this to activate the Digital components of the radio.

➔ LICENSE CLASS ABILITIES

Let's briefly discuss the limitations and abilities of the three license classes. We will not discuss the older two still in use by a few, Novice and Advanced. The reason is that you can never be granted these licenses. These two are the Novice (CW only) and the Advanced (falls between General and Extra).

We will discuss the **Technician**, the **General**, and the **Amateur Extra** license classes.

Let's discuss each in greater depth to understand the level of education and the communications capability you can expect.

MORSE CODE:

- ➢ No, it is not a requirement.
- ➢ No, you never need to use it.
- ➢ YES, you can learn it and use it as you wish.

These days, the requirement for CW or Morse Code has been removed from the requirements to be licensed. People learn it and use it these days for fun. There are plenty of ways to "cheat," like using a computer to send and receive CW; reading it is all you do. Typing your message and hitting send can make you look like a CW hero, but very few, if any, will do that regularly.

CW is the original data, being a binary means of communication. Ones and zeros. On and off. Beep and silent.

Any license level can use a repeater, any license level can use a DMR connection, and any license level can use simplex.

Another option for licensed operators is Echolink. It connects to specific repeaters and allows you to use your phone, radio, or computer to contact others anywhere on the planet.

EchoLink® software allows licensed Amateur Radio stations to communicate with one another over the Internet, using streaming audio technology.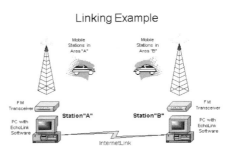
The program allows worldwide connections between stations or from computer to station, greatly enhancing Amateur Radio's communications capabilities. There are over 350,000 validated users worldwide — in 159 of the world's 193 nations — with about 6,000 online at any given time.

There are several ways to learn Morse code or CW, but the best way is to practice. Build a code practice oscillator.

https://w4cec.com/wp-content/uploads/2023/02/Popcicle-stick-CW-Practice-Oscillator-Project.pdf

Some classes and clubs can assist you in learning CW as quickly and thoroughly as possible. The ARRL sells audio CDs you can play in the car, allowing you to hear the code and pick it up faster. Hearing the letters is one thing, but hearing the words means you understand.

Check out the Long Island CW Club as they hold classes.

➔ A real online presence

One small aspect of amateur radio is that if someone has your call sign, they know who you are or can at least find your name.

https://QRZ.com is where everyone looks for the WHO of the new call sign.

Once licensed, it is a good idea to "CLAIM" your QRZ page and add information about yourself so others can learn who you are and what you do for amateur radio.

If they go to my QRZ page, it talks about my radio beginnings, clubs, repeaters I would be using, and any other activity I want people to know about me, like my writing and book sales.

You can put on that page whatever you want, within reason, of course. But suffice it to say, It is referred to as your "BRAG" page. A place where you can brag about YOU! As in your radio life, personal life, and other hobbies, if you sell things, put a link like I did for my books and my special discounts and deals that are only available IF you found them in this way.

It also contains a list of my equipment. Going to look at it when writing this paragraph, I realized it needed to be updated, as I have some new toys that I use and a few older toys I have given away to new HAMs.

The next page is that list. A list of my gear. If you see something interesting or have no idea what it is, GOOGLE IT! That is called research, and HAMS do a lot of that in their search for whatever they want to accomplish.

I wanted to research a little on finding a very light 2-meter antenna, and I found one! Made from copper foil and duct tape. It weighs a couple of ounces, and you can toss it into a tree to give you an excellent range.

My gear:
HAM SHACK and MOBILE:

ANALOG-HTs:	UV-5R, BF-F8HP, UV-82HP, UV-3R, Wouxun KG-UV2D
DMR-HT:	BTech 6x2
MOBILE:	Anytone AT-5888UV (mobile) ICOM IC-2820H (shack)
HF:	Yaesu FT-900 (semi-retired) Yaesu FT-891 (main radio for POTA and other remote operations)
ANTENNAs:	Tram dual-band Mag Mount (main mobile) Off-Center Fed Dipole cut for 40m EFHW (I purchased this kit from the ARRL) main --> *SB 1000 TIA vertical from Wolf River Coils* 2-Meter Slot Cube Antenna (needs repair) Several other random mag-mounts I give to people in need (mostly HT use- new techs - in cars) I made a few window wire dual-band roll-ups, also. KB4OI gave me a nice roll of ladder line and I will be making a bunch of 2m/70cm rollups to use the remote or to give away.
ARDF:	3-element Yagi (tape measure) with an Active Attenuator built on PVC tubing used for ARDF MicroFOX
SOLDERING:	I have a really nice soldering station. Eventually, the FCARC club wants to hold a soldering class. My station will be made available for this purpose.
SWR:	I own and use a NanoVNA-F. This is one awesome unit!
pseudo-Repeater:	2-HTs (UV-5R receive and BF-F8HP transmit) plus a cheap duplexer-type connection. It works in a clear waterproof case with a solar cell phone battery charger and will run continuously. It has 3800mAh batteries on the radios 2 SMA antennas connected to SMA 15-foot coax means it is high enough to connect as needed. This "repeater" operates on 145.000MHz (+) PL:100

➔ What can I do once licensed?

There are a lot of things you can do once you get licensed. For starters, join a club, maybe a few of them. They need your support, and they need your service as a volunteer.

You can enter Emergency Services and assist your local AUXCOM, ARES, or RACES organization.

Look for it online. They are easy to find and need licensed amateurs to maintain communications in the event of an emergency, natural disaster, or a practice to maintain a state of readiness in your area to help, assist, and protect your community.

Join local clubs and get information regarding the above and more. Join your club on FIELD DAY, an annual event where the club can practice disaster preparedness activity. Hence, they know that they are ready to help.

Personally, I am a member of several clubs:
1. ARRL
    a. American Radio Relay League
    b. https://arrl.org
2. FCARC
    a. Franklin County Amateur Radio Club
    b. https://www.fcarc.net/
3. OARS
    a. Occoneechee Amateur Radio Society
    b. https://facebook.com/groups/oars.radio/
4. TALARC

a. The American Legion Amateur Radio Club
   b. Wake Forest, NC American Legion, WF4TAL
5. JARS
   a. Johnston Amateur Radio Society
   b. https://JARS.net
6. RARS
   a. Raleigh Amateur Radio Society
   b. https://RARS.org
7. K2BSA
   a. National BSA (Scouting) Amateur Radio Club
   b. https://k2bsa.net/
8. FCHRE
   a. Five County HAM Radio Enthusiasts
   b. http://www.fivecountyhre.org/
9. Laurel VEC
   a. Volunteer Examiner group
   b. https://www.laurelvec.com/

Something else that is a great activity is to attend, or better yet volunteer at, a local HAM Fest. This is an excellent opportunity to meet new ham operators and get in a bit of networking, see and touch new equipment fresh on the market, talk to all the clubs and organizations in your area, and, of course, talk to and join the ARRL, and your local emergency services teams.

There are HamFests nationwide, and I attend a few of them yearly. In addition, I usually assist in any way I can because a volunteer organization needs volunteers to be a viable service to the community.

JARSFest in Benson, NC, and RARSFEST in Raleigh, NC, are some of them.

RARSFest is the day before (the Saturday) Easter. Check out **https://RARSFest.org** to volunteer your time and services. RARSFest is located at the North Carolina Fairgrounds.

JARSFest is a Sunday around the Thanksgiving timeframe. JARSFest is located at the American Legion in Benson, NC.

Most clubs have what is called a NET. A net is nothing more than a conversation over the air, but one difference. There is a person in charge of the net called the controller. That controller will ask for check-ins. You throw out your call sign, and the net control will contact the hams in order. When they get to you, you can say hello, tell what you have been up to in the world of radio, or whatever the theme of the moment is. Others will hear you; if you need something, you will likely find it quickly. They will ask Net Control to comment and speak to you directly if they have a question.

The net controller will also have a set of announcements so you can stay abreast of the activities you may be interested in in your area and with clubs OTHER than the club with which the net is associated.

It is not a competition in that sense. Still, most operators join multiple clubs because they believe in the message and want to help.

You have had a pretty good overview of amateur radio in the previous pages. Let's talk about each of the available licenses you can use.

➔ The *TECHNICIAN* class license

This is where you start. This is the entry-level license. The first exam you take is for the TECHNICIAN class license. This gives you the following:

a) The FCC Amateur Radio License (call sign)
b) Privileges on the 6-meter band (50 MHz), the 2-meter (VHF or 144 MHz), and the 70-centimeter (UHF or 440 MHz) bands
c) Privileges on a small portion of the 10-meter band for voice communications, 28.3 MHz to 28.5 MHz.

OK, suppose you study or have a lot of life experiences; perhaps you are an engineering student at a local college. In that case, you can take all 3 exams in one sitting. This would mean your "entry-level" would be Amateur Extra, and your privileges would be as such. This is very rare, but it does happen several times a year. This has happened only 4 times in all the exam sessions I have given.

Many Technician Class operators are happy with the ability of their licenses to operate on the VHF and UHF ham bands. However, technicians can also utilize many other technologies such as DMR, Digital Mobile Radio, or Echolink, the connectivity of a computer or phone that connects you to a 2-meter or 70-centimeter repeater anywhere in the world. All you need to do is apply, download the app, and go!

The 2-meter band goes from 144-148 MHz, and the 70-centimeter band goes from 420-450 MHz. These provide the Technician Class Operator with direct communications to another radio (called simplex) when you are near each other. You will also be able to utilize the abundance of repeaters, which will send your voice a great deal farther because the repeater is located a LOT higher and has considerably more power than your handheld or mobile unit. Most hams have dual-band radios – both HTs they carry on a belt and mobile radios mounted in their vehicles – that allow operation on both bands.

There are a few other bands that the Technician Class Operator is permitted to use, including the 6-meter band (50-54 MHz), the 1.25-meter band (222-225 MHz), and the 23-centimeter band (1240-1300 MHz). Yes, that is 1.3 GigaHertz. Radios in that range are expensive.

The 6-meter band sometimes allows extended range because of band conditions — where it may be possible to talk to other hams a great distance away on this band. The 6-meter band is called the "magic band" for this reason. In addition, it does some pretty unique and remarkable things entirely randomly and out of the ordinary.

Technicians can talk around the world, too!

Techs can operate on all VHF and UHF ham radio frequencies from the 6-meter band up through the millimeter-wave bands and on HF (high-frequency) bands. These bands provide worldwide communication, depending on current signal propagation.

Technicians can also communicate on HF on the 40-meter, 15-meter, and 10-meter bands operating in CW

(Morse code) mode. A small slice of those bands is set aside for CW, and Technicians are authorized to use it.

Technicians can use voice on the 10-meter band in the 28.3-28.5 MHz segment. There is plenty of room in that bandwidth when 10 meters is open (the band conditions are just right) to round up a bunch of contacts worldwide. Solar Cycle 25 is in its early stages, so the 10-meter band will open more often.

The 10-meter band is fun to work with, and Technicians can transmit up to 200 watts on 10-meters.

As a Technician, your primary goal should be to increase your license. Therefore, passing the General exam should be a priority. The exam is similar to the Technician exam, maybe more involved and granular. Still, it provides a monumental leap in capability and where you can transmit.

I tell my students to take pretests at **https://HamStudy.org,** and once you consistently pass with scores of 80 to 85 percent, look at the General class to get familiar with it. Then, when you pass the Technician test and take the General, you may just surprise yourself.

After all, you can miss 9 answers (out of 35 questions) on either test and still pass that element.

Let's talk about the **General Class** now.

➔ The _GENERAL_ class license

The General Class license gives you more privileges with a slightly more difficult FCC exam. However, many people will pass the Technician and the General in the same sitting since the information is essentially the same.

Once you know the Technician material, review the General material as well. You may just surprise yourself and pass the test. You are allowed to get 9 incorrect answers on the Technician and General exam. If you get 9 or fewer wrong, YOU PASS! Period.

The advantage of upgrading from the Technician to General class is the additional bands you can work with and an impressive power increase.

Also, the Technician is restricted to 200 watts on HF, whereas Generals can transmit up to 1,500 watts.

That's a lot of power! But remember, with immense power comes immense responsibility...... I mean cost.

Most HF radios transmit at 100 Watts, and a few transmit at 150 watts.

Any amplifier you get to increase above that will get you noticed, but remember that the cost is $2 or $3 per watt for an amplifier. Therefore, a 500-watt amplifier runs between $1,000 and $1,500.

Most Technicians I know who upgraded to General did so because they were tired of being unable to use the additional bands.

There is also the ability to get involved with Amateur Radio Emergency Services (ARES). A Technician can volunteer, but a General Class license will make you more valuable to the organization.

General class licensees can access all amateur operational bands, just not every amateur HF frequency. During disaster situations, amateur HF communications gets through without failure.

➔ The *AMATEUR EXTRA* class license

Extra class hams use all the same bands that General class licensees use on HF, but they get some additional segments within those bands. These segments are what make Extra Class an exclusive club with unique benefits! In addition, these additional frequency ranges make the Extra class license desirable to some hams.

There are 50 questions on the Amateur Extra FCC Exam, and you must get 13 or fewer wrong to pass.

So, if you are looking to do VHF and UHF with a little HF, a lot of HF, or HF with extra benefits, there is a ham license just for you.

The best thing is that you can stay at any class for as long as you wish before upgrading — if you even want to upgrade.

→ Which license class is best?

If you were happy with the limitation of using VHF and UHF radios, the Technician class license would be perfect for you.

On the other hand, handheld and mobile radios are readily available at reasonable prices, allowing you to connect with many hams in your local area, through Echolink, or perhaps over DMR, which uses the Internet to transmit your voice packets digitally to the destination you point it to anywhere on the surface of the Earth. I am looking for the day the ISS is on DMR through the TGIF network. Now, that would be cool!

Finally, suppose you want to talk worldwide and don't mind spending more money on HF radios and antennas. In that case, the General class license will be your fun zone. The general class also allows you to learn about and develop projects, experiment, build, and use antennas from your creative imagination. Yes, you can purchase assembled and tested antennas, whatever you need, but where's the fun in that?

As a Technician, you can build 2-meter antennas that are unique, interesting, and fun. The Slim Jim is the craziest. Clip, clip, solder, and POOF, antenna. I made my first a while ago and used it to contact a repeater a LOT farther away than I could reach with the stock antenna on my HT. So, I replaced it with my standard antenna and could not reach it. It's crazy, but that antenna cost me a few cents, and it is one of the best antennas I have in the woods for my HT.

Many antenna styles, shapes, and designs are out there. And looking around the web, you will quickly find that many plans are available for download. Some are better, some are worse, some are wild in design, and some are simplistic. It all depends on what you need or, better yet, what strikes your fancy!

As a General Class operator, you can use all available bands. Still, a small portion of the band(s) is reserved for the Amateur Extra class operator. You can listen on that portion, but if you are not an Extra, you cannot transmit into that portion.

Also, please remember that transmitting "close" to the end of either side of the band may violate the frequency restriction. Bandwidth is 3khz, after all. So don't accidentally violate the rules.

→ About the Author

Chris Cancilla was born in Cleveland, Ohio, on the East Side, in an Italian neighborhood called Collinwood, near East 158th and St. Clair. He really liked growing up there and would not trade it for anything. The friendships he made in Elementary School at Holy Redeemer and in High School at St. Joseph (now called Villa Angela – St. Joseph's) are priceless, and some are still in force. For most of his youth, he worked in the family business, DiLillo Brothers Dry Cleaners, for his Grandfather Carmen DiLillo, and at DiLillo Brothers Men's Wear for his uncle Tony (everyone called him the Czar). He also "apprenticed" with his Uncle Duke, an old-school radio and TV repair shop between the men's wear store and the dry cleaners. But he enjoyed working in the dry cleaners for his Grandfather the most. Two employees, Bertha and Evelyn, were like his second mothers.

In his youth, he really enjoyed Scouting. Spending a significant portion of it in multiple Cub Scout Packs, Boy Scout Troops, and Explorer Posts. Scouting influenced his life positively, and the training, knowledge, and education he gained during his youth in the troop still influenced his decisions as an adult. The ideals of Scouting, especially the Oath and Law, serve him today as a moral compass, guiding his actions to be a man his family can be proud of in all aspects of his life.

After high school, Chris spent 14 years in the US Air Force, where he saw a large chunk of this 3rd stone from our star. One of his favorite assignments was to Lowry Air Force Base in Denver, Colorado, where he could ride motorcycles and camp in the Rocky Mountains. This is a close second to the 2 years he was assigned to and lived in Keflavik, Iceland. He and his wife Tammy became best friends and experienced odd and unique landscapes and adventures. One was the SCUBA Diving Club's Founding President at Naval Air Station Keflavik. The name of the club was:

"vörn kafara á Íslandi"

He and his wife Tammy live in Raleigh, North Carolina, close to Wake Forest. He really misses his little buddy and writing partner, his cat, Snip. Snip followed Chris

around from room to room. You may or may not see him all the time, but he is always close by. Unfortunately, snip crossed the rainbow bridge a couple of years ago; he went fast, which is the only consolation. When Chris writes, though, he still is close by. They made a paw print before he was cremated, and that paw print always sits on the desk near the computer.

The Boy Scouts of America is still a part of his life, especially in teaching new adults the skills needed to survive the outdoors and reinforcing how these outdoor skills and habits need to be introduced to the leaders of tomorrow. Leave No Trace camping is a significant part of his instruction and is a philosophy in the conservative style of camping Chris enjoys, if not the only way to ensure an excellent time for you and future campers. Wilderness camping is a great way to decompress and gain insight into what is hidden in the inner recesses of your mind. Sitting around a campfire on a cool or cold night, watching the flames dance, and watching the wood that has given its all to the moment's beauty allows you to reflect on your thoughts and be honest with yourself. The one person you cannot lie to is yourself, so honesty in your head provides nature to clarify all things.

Imagine you are asleep for a moment, and a noise wakes you. You realize you left the Dutch Oven on a picnic table, thinking you would clean it in the morning. Well, you spend the next few hours arguing with a 50-pound raccoon about the cobbler residue in the Dutch Oven on that picnic table, the same Dutch Oven you said you would clean up in the morning. Sometimes, you let the raccoon win!

Chris also has a passion for cooking. Creating several cookbooks allows him to experience new cuisines and cooking methods from around the globe. Still, it also gives him the ability and materials to share and teach cooking to less experienced or knowledgeable people. He does not consider himself a chef, but he does consider himself a somewhat OK cook, both in the home and in the woods.

Cooking in the woods is a skill that not all that many people have even considered. However, it is one skill that Chris enjoys teaching to Scout Leaders, both old and new, in classes he teaches for Scouters (Adult Boy Scout Leaders) and the Scouts themselves during the COOKING Merit Badge. Chris was happy that the BSA finally made cooking a required merit badge for the Eagle

Scout rank. It is a skill that will be valuable for the rest of your life. Especially if you want to prepare a romantic meal for a date or simply provide a meal you enjoy.

Whenever Chris develops or finishes a new story or cookbook, he permits some people to read his book and offer ideas to improve the storyline or the text. In addition, he may allow you to be the next editor, for which he will give you kudos at the beginning of the book. Thus immortalizing you in the story for all eternity.

His last hobby is Amateur Radio. In the Raleigh, NC area, you can find him in the mornings on K4ITL and in the evenings on AA4RV; he pops in occasionally to AK4H. If you use a DMR (Digital Mobile Radio), try to make a QSO with him on the TGIF Network, Talk Group 1870. He usually monitors that talk group and would enjoy the QSO.

I hope you enjoyed reading this book. Please read others in the series or check out the cookbooks or both if you are interested in cooking. Also, pick up that briefing booklet if you work with an EDI team and do not understand Electronic Data Interchange. It is well worth your time to read. Reviews of those who previously read the book are in. It is a well-received and informative book that can bring someone to understand EDI's fantastic and fun world. Tell Chris what you think of the books you read and whether you liked the stories, the briefing, or the recipes.

Chris's day job is as an EDI B2B Integration Specialist or an EDI Developer. Take your pick; they both mean the same thing. He calls himself a digital mailman. He moves the data and information files from one place to another. Still, he does not own, nor is he responsible for, the data in any way other than delivering it. ☐ So, a mailman!

That's a fancy way to tell someone you work with computers to translate data from one format to another. After all, the mailman doesn't write the letters but moves them from point A to B.

To find more information about Chris, check out

https://AuthorCancilla.com

you can order any currently available books he has created here.

For more information regarding Amateur Radio,

please go to

https://W4CEC.com

You can find classes, exams, and projects to make your radio life more enjoyable.

Additional Works by Christopher E. Cancilla

All these titles are available at:

https://AuthorCancilla.com

The Archives, a 7-Part, Time Travel Novel Series

Revised, edited, and renewed as of October of 2023

1. The Archives: Part 1 – Education
2. The Archives: Part 2 – Fixing Time
3. The Archives: Part 3 – Salvation
4. The Archives: Part 4 – Family
5. The Archives: Part 5 – Fresh Start
6. The Archives: Part 6 – Continuum
7. The Archives: Part 7 – Temporal Logs

EDI Education Series, a 5-Part briefing providing an understanding of EDI

1. EDI Education: Briefing 1 – Introduction – What is EDI, and how does it work? Read and Learn!
2. EDI Education: Briefing 2 – Deep Dive – A Deeper Dive into the 850/Purchase Order
3. EDI Education: Briefing 3 – Getting Paid – a Deeper understanding of the 810/Invoice
4. EDI Education: Briefing 4 – Shipping – Demystifying the 856/Advance Ship Notice
5. EDI Education: Briefing 5 – The Complete Briefing – A review of the first 4 books with additional insight

Free to Read Stories

available on http://AuthorCancilla.com

1. Stargate Universe
2. Scorpion Sting
3. Terra Nova

Additional Science Fiction books available

1. The Ultimate Thru-Hike
2. Bus Route 40-A
3. Lost Earth
4. Colony 3
5. Mountain Life

Other available books and novels

1. AMMO – IYAAYAS
2. Toasting Marshmallows on my Dumpster Fire
3. Getting Published
4. Getting Published Two
5. Life as an Amateur
6. Stories from Time and Space
7. Scouting and Camping: A New Parents Guide
8. Scouting to Summer Camp
9. Camp Menu Planning
10. Personal Menu Planning
11. Learning to Camp
12. Packing your Backpack for a 5-Day Trip

Discounts and Deals

For current and available discounts, go to
http://AuthorCancilla.com

1. ARCHIVE: the 7-part series
2. EDI: The Complete Briefing
3. Learning to Camp and Learning to Backpack

Made in the USA
Columbia, SC
28 March 2024

33654910R00043